THE SPOOKY WORLD OF QUANTUM PHYSICS

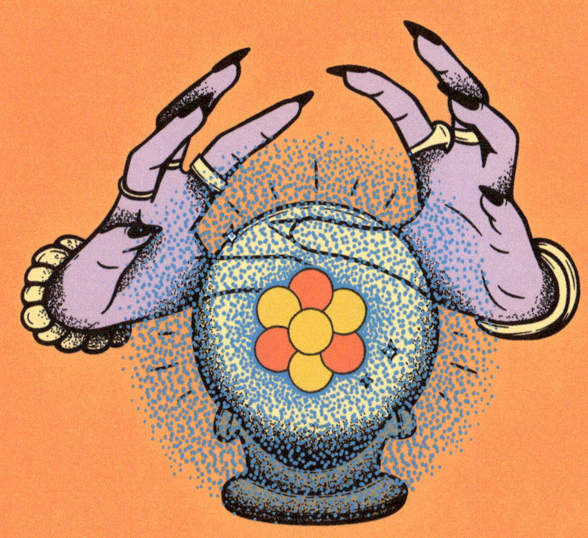

PREETINDER RAHIL

The author may have checked for errors twice or even thrice, but doing your due diligence would still be considered wise.

NOTICE

Missing Electron

Negatively charged and
extremely small.
Please contact Werner
Heisenberg if found.

where could it be?

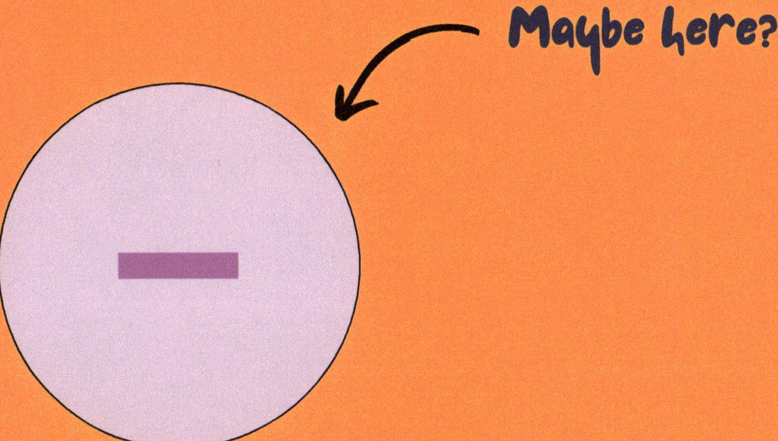
Maybe here?

Maybe there?

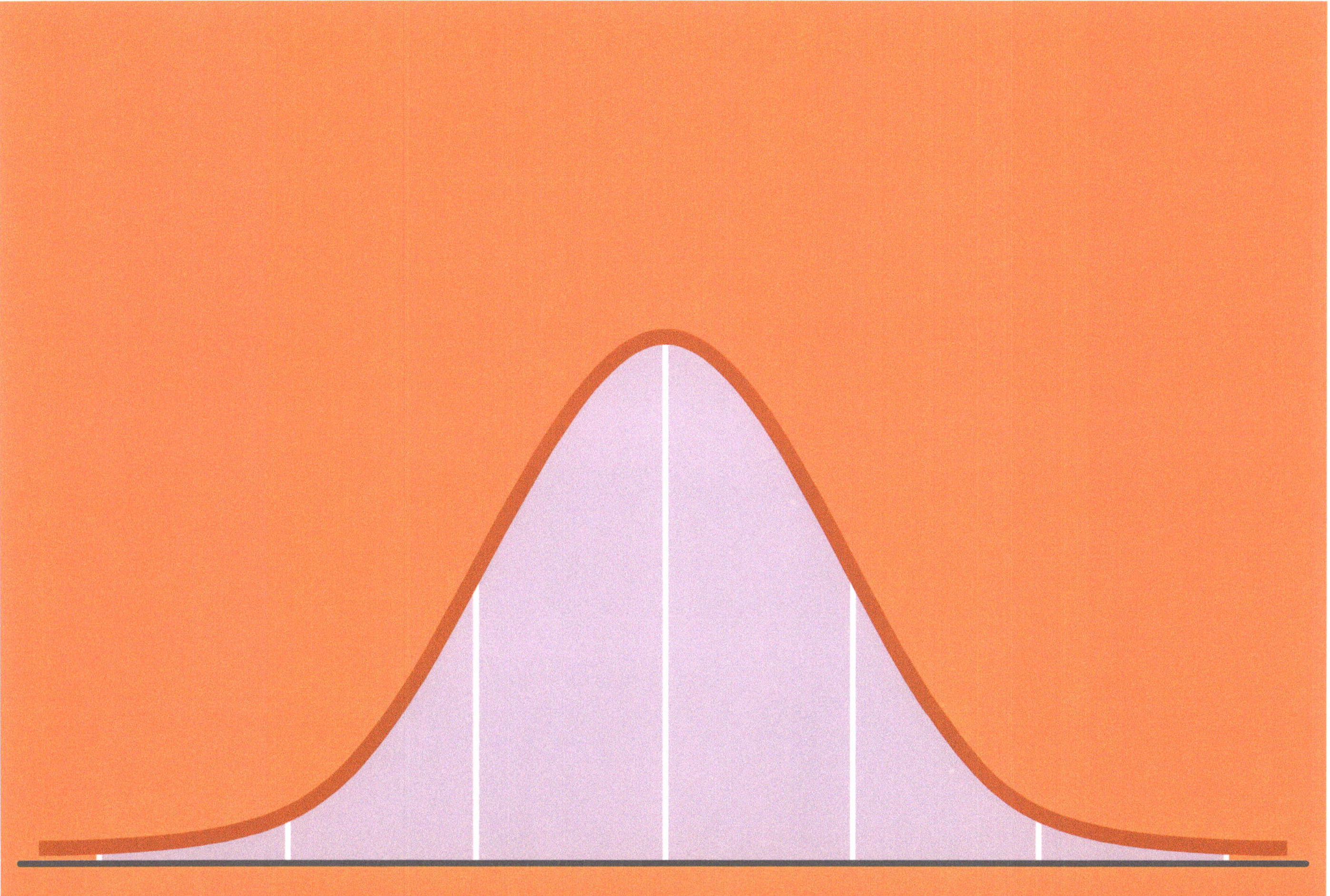

it could be anywhere and everywhere.

its position is fuzzy unless we decide to look.

Does something exist only if i look at it?

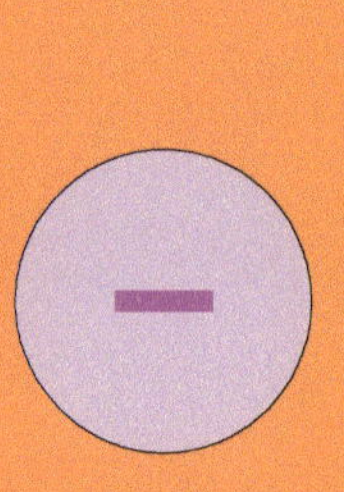

When we look, the electron suddenly appears.

But it appears at a
random place as if we are
rolling a dice.

THAT'S

SPOOKY

Even if we know where the
electron is, we still don't
know where it's going.

Heisenberg called it the uncertainty principle, but we'll call it

is your head spinning yet?
That's perfect timing to
learn about electron spin.

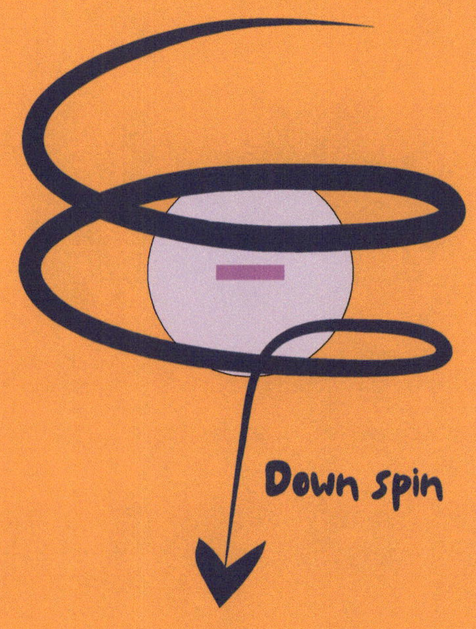

The electron can spin.
it can either have up or down spin
but only when we measure it.

But unlike a spinning
basketball, the electron
cannot be made to spin
slower or faster.

The electron will never stop spinning as it needs no effort to spin.

This is how physicists react when asked what causes an electron to spin?

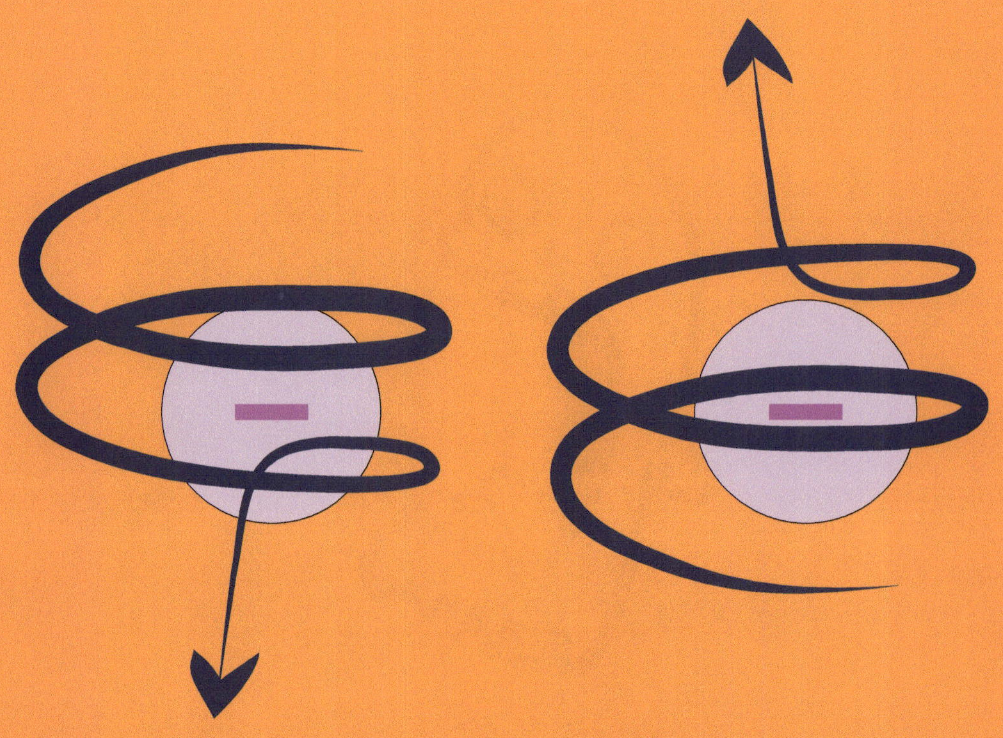

Two electron spins can become entangled. If one's spin is up, the other must be down.

But before the measurement electron spins are not set, they remain fuzzy.

The act of measurement is like magic. it forces the electrons to choose their spins.

Even if we take the electrons to the edges of the universe, they can still communicate instantly.

Even Einstein called it spooky action at a distance.

Then there are so-called ghost particles called neutrinos that can pass through our body without us noticing at all.

Our universe is heavy thanks to the dark matter. But we still remain in the dark about this matter.

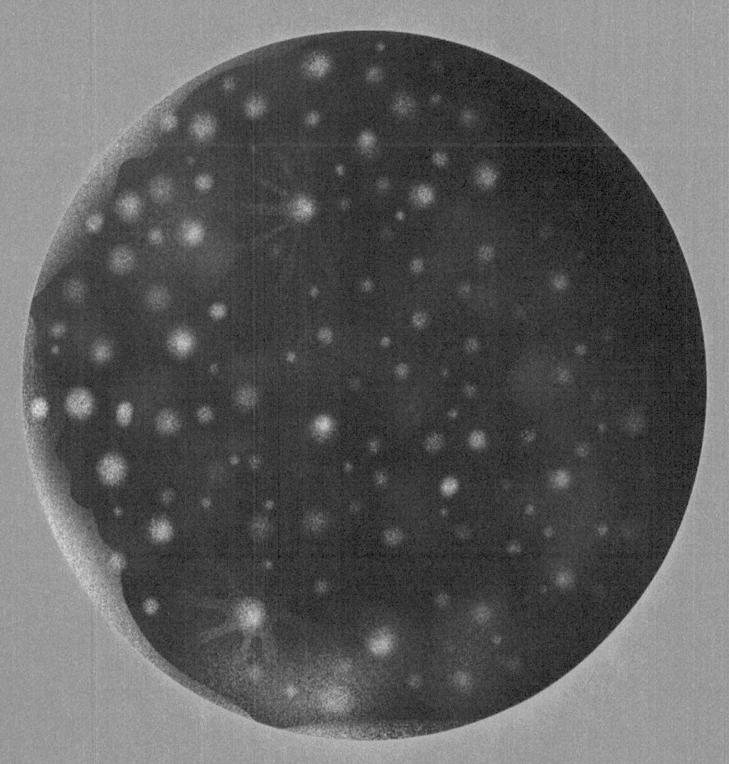

we can't see the dark
matter, only feel it. is't it
spooky?

Then there's dark energy which is pulling the whole universe apart.

The fate of the universe is at stake. Will we end up with a cold dark universe?

Reality is way more spookier than made-up ghost stories. Do you agree?

About the Author

Preetinder Rahil writes fiction, non-fiction, and poems that rhyme. He tries to keep things simple, fun, and worth your time.